"When I consider Your heavens, the work of Your fingers, the moon and the stars, which You have set in place, what is mankind that You are mindful of them, human beings that You care for them?" Psalm 8:3-4

Using this book

All *italicized* words in this study guide match the 1st to 3rd grade lapbook pieces. Depending on your child's reading level, grade, or ability, you may choose to: read only the italicized portions to walk your youngest child through the lapbook, allow the oldest students to read the book and work independently, or anywhere in between. If the child is able, he may choose to read the book independently and fill in the accompanying pieces. The illustrations in this study guide also match those found in the lapbook to aid the student in locating information. If the child is on a particular lapbook piece, and does not know the answer, he need only to find the matching picture in the study guide and the answer will be close by (for 4th-6th+) or *italicized* (for 1st to 3rd). You may also choose to read the study guide in sections to your students as a group, and then assign them to work independently at their level in the lapbooks. This can be done by segmenting readings. For example, reading a planet a day. Or, parent may choose to assign a section a day. For example:

1st grade: parent/teacher reads italicized portion of "Space Rocks" to the student, then provides the word bank and lapbook pieces for "Space Rocks". Student works with assistance to fill in correct answers.

2nd grade: parent/teacher or student reads italicized portion of "Space Rocks", then provides the word bank (if needed) and lapbook pieces for "Space Rocks". Student works independently to fill in correct answers.

3rd grade: student reads "Space Rocks" independently, and using the information in the study guide (no word bank), fills in the lapbook pieces.

4th-6th grade: student reads "Space Rocks" independently then answers the written questions next to each lapbook piece in complete sentences.

All students: Parent reads "Space Rocks" section to all students simultaneously, and then assigns students the grade appropriate lapbook pieces for independent work, helping the youngest students as needed. There is also an index at the back of the study guide to further assist in locating information.

I hope you enjoy working through this study guide and the available lapbooks with your children as much as I did mine! HAVE FUN!

1st – 3rd grade lapbook:	4th – 6th grade lapbook:
ISBN-13: 978-1478204398	ISBN-13: 978-1478204459
ISBN-10: 1478204397	ISBN-10: 1478204451

Chapter 1: Vocabulary

astrology- the false belief that the stars, planets, and universe hold power over us and our future.

astronomer- a person who studies planets, stars, and outer space.

astronomy- the study of the planets, stars, and outer space.

atmosphere- the gases that surround a planet.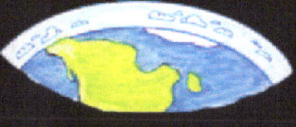

axis- the imaginary line that goes from the north pole to the south pole.

celestial- any natural object in space such as planets, asteroids, or comets.

constellations- the pictures the stars appear to make. They are often used for navigation.

equator- the imaginary line that divides the earth in half.

experiment- a scientific test to discover something new.

gravity- the invisible force that draws two [celestial] objects together. Sir Isaac Newton discovered gravity.

heavens- the word the Bible uses to describe everything God created outside of Earth.

hemisphere- the two halves of Earth divided by the equator.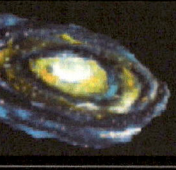

magnetic field- the area around Earth that has a magnetic charge caused by the core of the planet.

moon- a piece of space rock that orbits a planet. It can be round in shape or an irregular asteroid-like object.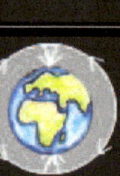

planet- an object that orbits a star, is round, and has cleared its path.

revolve- to orbit or go around another object in space such as a star or planet.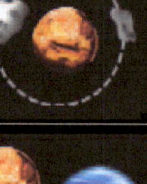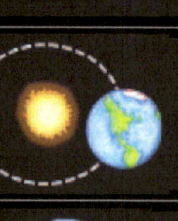

rotate- to turn or spin on the axis or center

satellite- an object that orbits a planet (a moon) or an object launched from earth that orbits a planet and transmits a signal back to earth.

space- everything beyond the planet Earth. (the heavens)

solar system- the Earth and planets that orbit our sun.

tides- the rise and fall of the oceans due to the moon's gravitational pull on the earth.

Chapter 2

Space Rocks and Comets

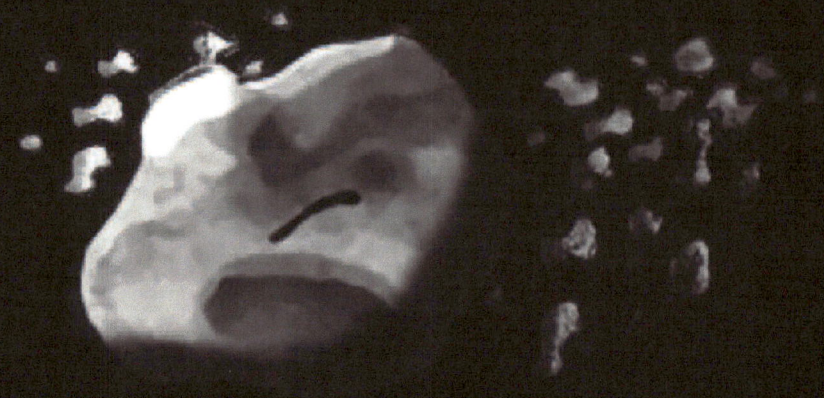

Space Rocks

Asteroid, meteoroid, meteor, meteorite . . . What's the difference?

Asteroid

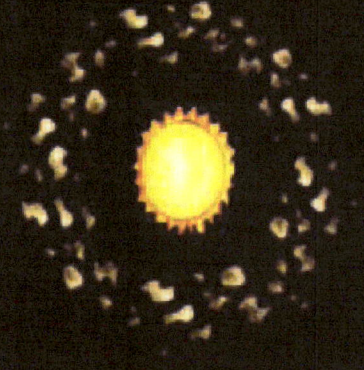

An asteroid is a piece of rock and metal that orbits the sun. They can be as small as 150 feet across or as big as a small planet. The main difference between asteroids and meteoroids is their size and movement. While a meteoroid drifts randomly in space, an asteroid orbits the sun. The majority of asteroids in our solar system are in the asteroid belt between Mars and Jupiter. The

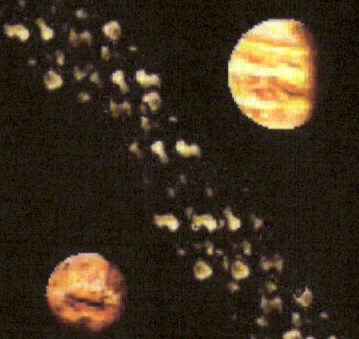

largest asteroid in this belt is Ceres at about 640 miles across. It is so large it is considered a dwarf planet and is the only known dwarf planet in this asteroid belt. One asteroid, Ida, is large enough to have a moon orbiting it.

The asteroids in this belt are not always very close together. Most of the asteroid belt is empty space with tens of thousands of miles between asteroids. There may be millions of asteroids; however, if all the asteroids were put together, it would weigh less than the moon.

Another asteroid region in our solar system is beyond Neptune and called the Kuiper Belt. Pluto, Charon, and Eris are dwarf planets in the Kuiper Belt that orbit the sun. There is not much information about the Kuiper Belt because it is so far away.

Meteoroid

A meteoroid is a small piece of dust or rock floating in space. They often have no orbit and drift randomly. Many scientists believe they form when two asteroids collide. A meteoroid is also any space debris that is smaller than an asteroid.

Meteor

A meteor is a meteoroid that has entered the earth's atmosphere. It is also called a shooting star or a falling star. Every day, dust and rocks are hitting our planet at about 158,000 miles per hour. Because of their size and speed, most burn up in the earth's atmosphere without us even knowing it. Sometimes we can see it as it burns up as a bright streak of light across the sky. This is when we call it a shooting or falling star. It is not really a star we see but a meteor.

Occasionally, thousands of meteors will streak through the sky. This is called a meteor shower. Meteor showers occur when the earth travels through the particles left from a passing comet. They are often easily predictable as astronomers observe a comet and the path of the earth in space.

Meteorite

When a meteor survives the atmosphere and lands on earth, it is called a meteorite. They are typically made of iron and stone. *The biggest meteorite ever found was discovered in Grootfontein, Namibia in southern Africa and weighs 66 tons.* It is called Hoba after the farm where it was found.

The largest meteorite found in the United States is 15 ½ tons and was discovered in the state of Oregon.

Sometimes a meteorite can make a large crater though it is very rare. There are 175 known craters on Earth and several are in the United States. Arguably, the best known crater is Meteor Crater (also called Barringer Crater) near Flagstaff, Arizona. Because it is in the desert, there is very little water to erode it so it is very well preserved. It is almost a mile across and over 550 feet deep. *The biggest crater known in the world today is Vredefort Crater in South Africa*. It is about 175 miles in diameter.

Comets

A comet is a ball of ice, dust, and gas that orbits the sun. Because of all the ice and dust, they are often called dirty snowballs.

The "ball" at the front of a comet is called the coma and its center is the nucleus. The nucleus is the icy core of the comet. Comets can range in size from one mile across at the nucleus to over 50 miles at the nucleus. When the comet nears the sun, it begins to heat up and the coma appears as the nucleus releases gas and dust. Then, two tails appear. *One tail is dust and the other is gas.*

These tails can be tens of millions of miles long. Sometimes if the comet is weak, only one tail or a glowing coma will be visible.

Comets appear to glow due to the sun and its interaction with the gas, ice, and dust that are released as the comet heats up. The light from the sun reflects off the dust tail and makes it appear to glow. It is the second tail which truly glows. The particles from the sun react with the charged gases escaping the comet and make the tail glow. Due to solar wind, *the tails of a comet will always point away from the sun*. When the comet has no more gas or ice to lose, it becomes a "dead comet".

Comets have an elliptical orbit around the sun that can take anywhere from 3 to 1,000 years to complete. There are so many comets that five or six can be seen every year depending on location. Only two or three of these are known comets, the rest are new ones never before seen from Earth.

Chapter 3

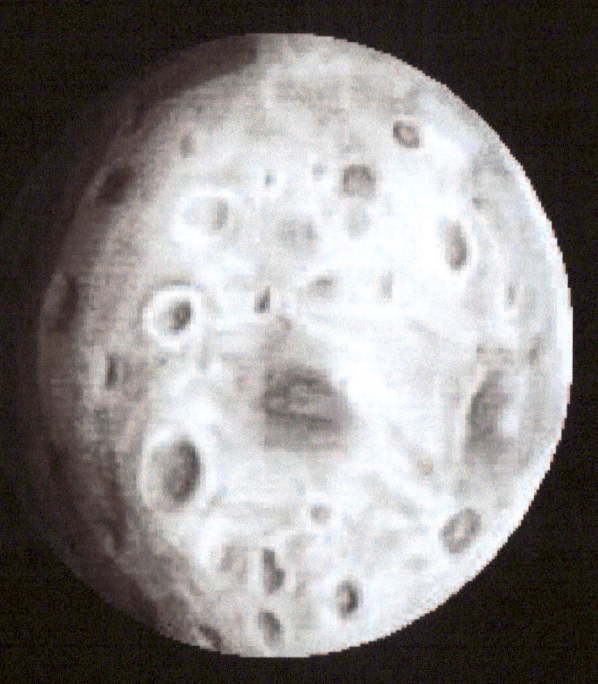

Mercury

Mercury is the first planet from the sun. It is also the smallest now that Pluto is a dwarf planet. Despite being the closest planet to the sun, it is still an average of 36,000,000 miles away. Mercury is unlike other planets in that its orbit is very elliptical (oval) and off center. This means the planet can be as close as 29,000,000 miles to the sun or as far away as 43,000,000 miles. Mercury has no moons.

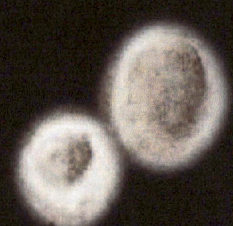

Mercury is a rocky, dusty planet with lots of craters. The Caloris Basin, at over 800 miles across, is the largest. Most of the craters on Mercury are named after famous poets, writers, musicians, composers, and artists like Mozart, Tchaikovsky, Picasso, Robert Burns, and Charles Dickens.

Mercury also has ridges and cliffs making it look wrinkled. Some scientists believe this wrinkling is because the planet shrank after it was formed.

The first planet

Size: 9,525 miles in circumference 3,025 miles in diameter

Distance from sun: average 36,000,000 miles

Origin of name: messenger of the Roman gods.

Moons: none

Average temperature: 350 ° F

Weight: 38 lbs. (If 100 pounds on Earth)

Mercury does not have an atmosphere and looks a lot like our moon. Because there is no atmosphere, Mercury becomes very hot in the day and very cold at night. It can be over 800°F in the sun, and as cold as -300°F on the side facing away from the sun.

Mercury orbits the sun faster than any other planet. It only takes 88 days to orbit the sun and it takes 59 earth-days to rotate once on its axis. Because Mercury rotates almost as quickly as it revolves around the sun, it actually takes 176 days to complete one day, sunrise to sunrise, at the same point on the planet.

 While we can see some planets from Earth easily, we cannot see Mercury well because it is too close to the sun. It can only be seen just before dawn or just after sunset along the horizon.

Venus

Venus is the second planet from the sun and has no moons. It has an extremely dense atmosphere of carbon dioxide and nitrogen that is 96 times greater than Earth's atmosphere. Venus is also the planet that is closest to Earth in size with a difference of only about 1,000 miles in circumference. Venus is 67,000,000 miles away from the sun.

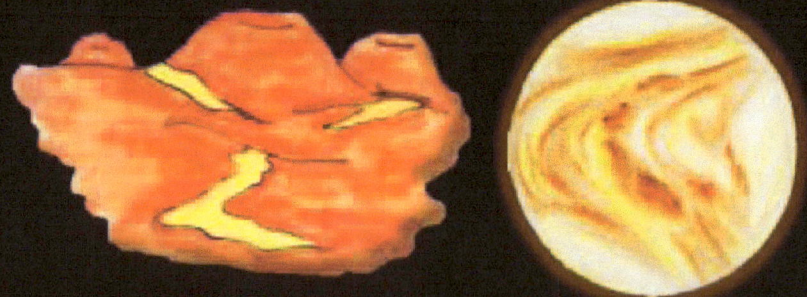

Space probes have revealed that Venus has a varied terrain of mountains, valleys, and volcanoes. It is covered in thick, yellow clouds made of sulfuric acid which reflect the sun's light.

The second planet

Size: 23,628 miles in circumference 7,521 miles in diameter

Distance from sun: 67,000,000 miles

Origin of name: the Roman goddess of beauty

Moons: none

Average temperature: 847 ° F

Weight: 90 lbs. (If 100 pounds on Earth)

Besides the sun and moon, Venus is the brightest celestial object in the sky. It is so bright we often see it as the first "star" of the evening, even though it is really a planet. During certain times of the year, it is the last "star" to fade just before sunrise. That is why *Venus is called the evening star and the morning star.* This is when we can see it so clearly.

Besides reflecting the sun's light to make it so bright, V*enus's thick yellow clouds trap the sun's heat making it the hottest planet. It can reach up to 890°F!*

Venus has a very interesting rotation. *Venus is the only one to rotate in the opposite direction (backwards) of all the other planets.* Some scientists believe the universe is billions of years old and that it was created when some gas and dust swirled around a new sun. Over time this gas and dust formed planets. This is called a theory. The Bible says that God created the heavens and the earth in the beginning. If you were to pull the plug on a kitchen sink full of water, all the water, bubbles, and leftover crumbs would swirl in the same direction around the center. This is similar to how some people believe our solar system formed. If that was true, why does Venus spin in the opposite direction? Scientists cannot explain why Venus does this, but it is clear that God created it this way!

Venus is also odd in that its rotation time is longer than its orbit. This means that its year is shorter than its day. *It takes 225 earth-days to orbit the sun and 243 days to rotate once on its axis.*

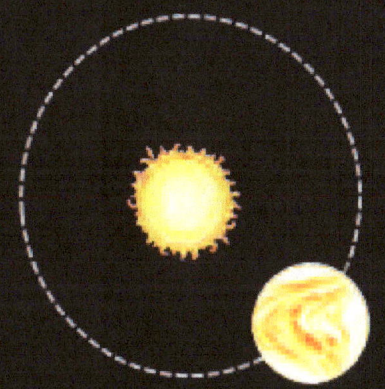

Earth

Earth is the 3rd planet from the sun and the only one with life. It is 93,000,000 miles from the sun. If it was any closer, it would be too hot and everything would burn up. If it was any farther away, it would be too cold and everything would freeze. We would have no liquid water which is essential for all living things.

The earth is always moving. It spins at almost 1,040 miles per hour. We cannot feel this movement because we are traveling right along with it. The earth turns or rotates on its axis once every 24 hours. The axis is the imaginary line that goes from the North Pole to the South Pole.

The third planet

Size: 24,900 miles in circumference at the equator 7,926 miles in diameter

Distance from sun: 93,000,000 miles

Origin of name: Anglo-Saxon word for ground

Moons: 1

Average temperature: 60 ° F

Weight: 100 lbs.

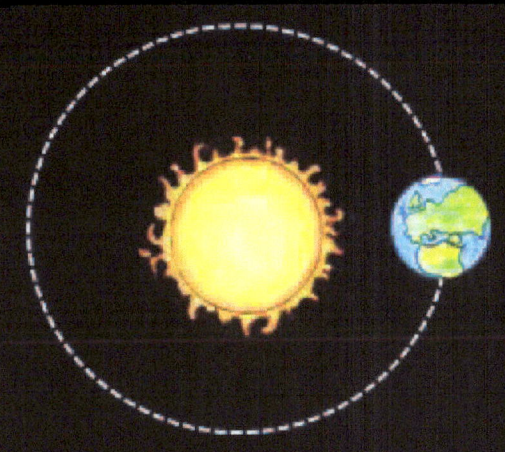

It takes one year or 365¼ days to revolve or go around the sun. Every four years we have an extra day to make up for the extra ¼ of a day it takes to orbit the sun. This is called a leap year.

In the first century BC, Julius Caesar and the Roman astronomers discovered that their Roman calendar of 355 days was not matching up with the seasons. Their year was passing more quickly than the seasons were. The astronomers decided to create a 365 day calendar with an extra day added every fourth year to keep the calendar in sync. However, over time, the seasons and dates were again out of sync. This time, they were ahead by 11 days. In 1582, Pope Gregory XII and his astronomers re-adjusted the calendar and changed the leap year. It would again occur every four years however only on a century year which was divisible by four hundred. Finally, the calendar matched the seasons.

When the world rotates, part of the earth is in shadow. In the shadow it is night, and in the sunlight it is day. Because of the Earth's tilt, the amount of sunlight we receive changes according to the season. The farther north one travels, the longer the summer day is and the shorter the   winter day is. Alaska is known as the "Land of the Midnight Sun "because during the three months of summer, the sun never fully sets! Of course, just the opposite is true in winter; the sun stays hidden for three months!

> Psalm 74:16 The day is yours, and yours also the night; You established the sun and moon.

During the day, it looks like the sun is moving across the sky, but it is really the earth moving. It travels through space at around 67,000 miles per hour.

The earth is not straight up and down; it is tilted on its axis at about a 23° angle. *This tilt gives us seasons. Without this tilt, the earth would get too hot by the equator and too cold at the poles.* Not only this, but it would throw our whole weather pattern off. The weather would remain somewhat stationary in a constant pattern. The temperatures would even out and each area would have a consistent temperature. The equator would be unbearably hot and only some of the land would actually be able to produce crops. God perfectly designed the earth with this tilt for us so that more of our land could produce food.

When the earth orbits the sun, the sun's rays shine more directly on different parts of the world and glances off other parts of the world.

winter   spring   summer   fall

Genesis 8:22 As long as the earth endures, seedtime and harvest, cold and heat, summer and winter, day and night will never cease.

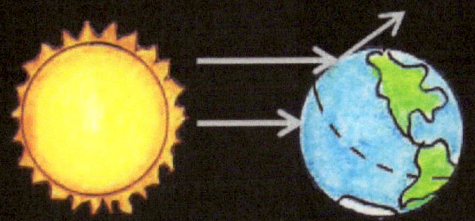
In the winter months, the northern hemisphere is tilted away from the sun, deflecting the sun's rays.

In the summer months, the northern hemisphere is tilted towards the sun allowing more of the sun's rays to directly enter the atmosphere.

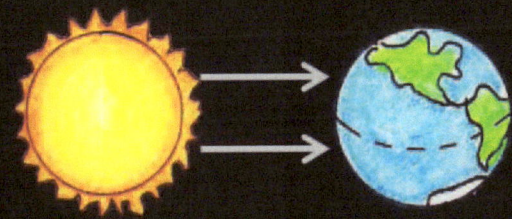
In the spring and fall, the rays are evenly distributed since earth is tilted to the side.

When it is winter in the northern hemisphere, it is summer in the southern hemisphere. This means that Christmas in the southern hemisphere is in the middle of summer! Likewise, when it is spring in the northern hemisphere, it is fall in the southern hemisphere.

While earth is just right for us, it does get very hot in some places and very cold in others. It can be as cold as -129°F in the Antarctic ice sheet and as hot as 134°F in Death Valley, California. Overall, our average temperature on the planet is about 60°F. It is quite comfortable for life!

We stay at such a perfect temperature because of our atmosphere. This is the air we breathe. It acts like a blanket trapping some of the sun's heat while protecting us from getting too cold. It keeps our temperatures consistent. We stay cooler in the day and warmer at night. Without the atmosphere, the difference in temperatures between night and day would be extreme.

Our world is perfect for life. Everything in heaven and earth is different, there is nothing the same. God has clearly shown His existence and majesty in His creation and the complexity of life. From the tiniest ant, to the polar bear, to the giant dinosaur, and then to man, God's signature is written for us all to see.

Romans 1:20 "For since the creation of the world God's invisible qualities—his eternal power and divine nature—have been clearly seen, being understood from what has been made, so that people are without excuse."

Psalm 104:24 How many are your works, LORD! In wisdom you made them all; the earth is full of your creatures.

Inside the earth are different layers. The three main layers are:

The crust is where we live. It is the coolest and thinnest part of the planet. If the earth was an apple, the crust would be as thin as the skin. The mantle is the thickest part of the planet. It is thick, hot rock. The center or core of the earth is the hottest and made mostly of iron. Each layer of the earth is hotter and denser than the one before it.

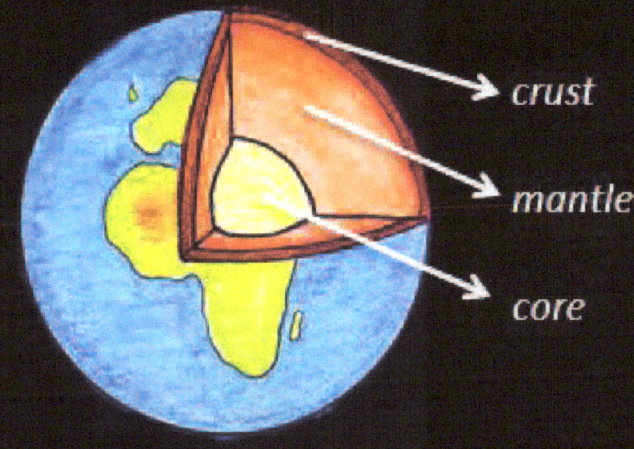

crust
- mostly basalt and granite; also silicon, aluminum, calcium, sodium and potassium.
- average thickness: 5 miles (under the oceans) to 25 miles thick (on continents)
- average temperature about 60 °F

mantle (upper and lower)
- believed to be magnesium, silicon, iron and oxygen
- begins about 18 miles down; 1,800 miles thick
- believed to be between 1,600°F to 4,000°F

> Hebrews 1:10 He also says, "In the beginning, O Lord, You laid the foundations of the earth, and the heavens are the work of Your hands."

core (inner and outer)
- outer core believed to be molten nickel and iron
  inner core believed to be solid iron (solid from pressure)
- outer core begins 1,800 miles down, 1,400 miles thick;
  inner core 1,500 miles in diameter (across)
- outer core believed to be between 7,200°F to 9,000°F
  inner core believed to be between 9,000°F and 13,000°F

The crust is broken into pieces called plates. Sometimes magma (thick, hot rock) pushes up between the plates and forms volcanoes. The magma is called lava when it comes out of the earth.

Under the earth's crust are pockets of magma. It collects in a magma chamber and often works its way out onto the earth's surface. Magma is very hot. Its temperature ranges between 1,300°F and 2,400°F. Most volcanoes in the world form where two or more plates meet. The biggest volcanic region in the world surrounds the Pacific Ocean and is called the Ring of Fire.

Sometimes the plates move, which causes the ground to shake. This is called an earthquake. They often happen where two or more plates meet. This spot is called a fault. During an earthquake, the fault can pull apart, slide back and forth, sink down, or rise up.

Earthquakes are measured using a scale called the Richter scale. With each higher whole number, the earthquake is ten times stronger than the number before.

There have been accounts of earthquakes in the Bible. When Christ died on the cross, the curtain in the temple tore in two, the earth shook, and the rocks split. The Bible also speaks of the mountains trembling before God.

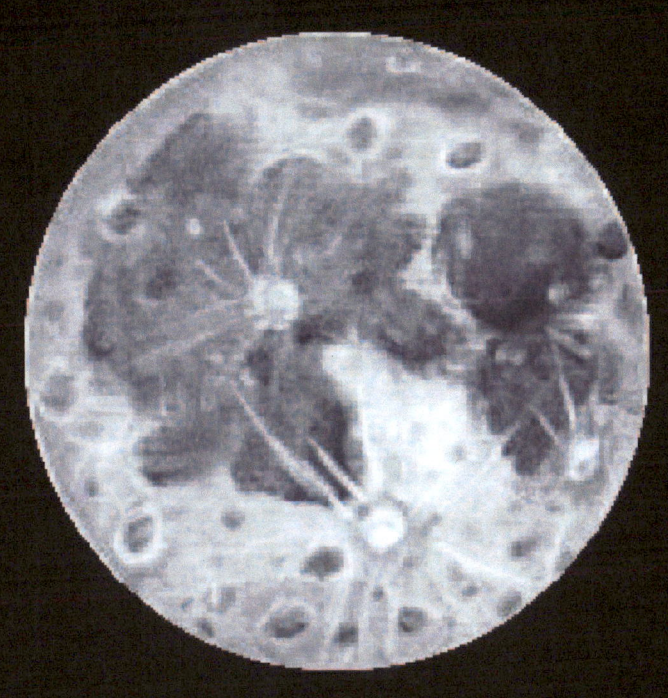

Moon

Many planets have moons. Moons are bits of rock that orbit or go around other planets. While Jupiter has the most moons, Earth only has one. The earth's moon is different from other moons because of its size relative to the planet it orbits. The moon is roughly 27% the size of Earth. The other moons in the solar system are significantly smaller than the planet they orbit.

The moon is also relatively close to earth, only 239,000 miles away. Scientists are now able to measure the distance to the moon to within millimeters thanks to the mirrors the Apollo astronauts left on the moon. Scientist use a laser, reflect it off these mirrors, and time how long

The moon

Size: 6,800 miles in circumference at the equator 2,160 miles in diameter

Distance from earth: 239,000 miles

Origin of name: mona- Anglo-Saxon for moon.

Average temperature: -10 ° F

Weight: 16 lbs.
(If 100 lbs. on Earth)

it takes to return to Earth. They have come to realize that the moon is actually receding (going away from) the Earth at a rate of about one inch per year. If the solar system was billions of years old as some theories say, the moon would not be orbiting us right now. It would have floated off into space! Or, if the moon's path was traced back billions of years, the moon would not be able to orbit as it would be sitting on earth! This is yet more evidence that the solar system and universe is much younger than some believe. This closely follows the Biblical timing of creation.

A day on the moon is about 27 days long and it takes 27 days to orbit the Earth. Day and night each last for two weeks! Because the moon's rotation and orbit are the same length, we only see one side of the moon from Earth. The other side of the moon is often called "the dark side of the moon" for this reason.

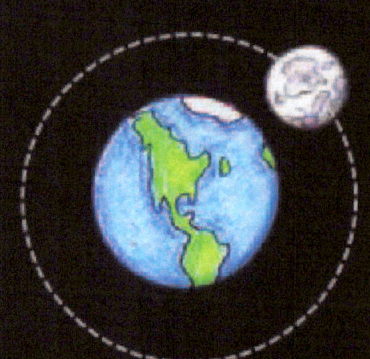

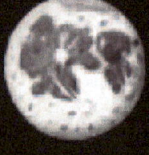

The moon is covered in craters and dark areas early astronomers believed were seas. These dark areas are actually old lava flows.

The moon seems to glow at night, but it does not make its own light. It reflects light from the sun like a mirror.

God gave us the moon so we could see at night.

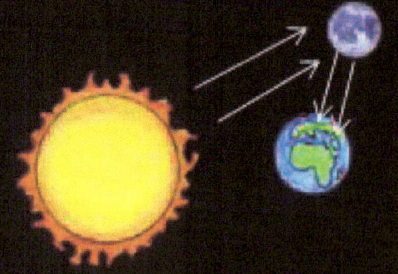

Genesis 1:15-16 And let them be for lights in the firmament of heaven to give light upon the earth; and it was so. And God made two great lights; the greater light to rule the day, and the lesser light to rule the night; He made the stars also.

There is no water, air, or life on the moon. Like Mercury, the moon does not have an atmosphere so there is nothing to protect it from the heat of the sun or the cold of space. It can be 260°F in the sun and -280°F on the dark side of the moon.

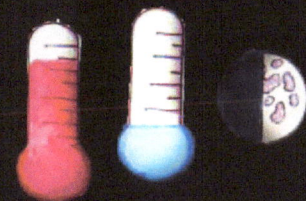

As the moon orbits Earth, we see part of it in shadow. Its shadow shifts and changes, moving across the surface as it rotates the earth. *These movements are called phases.*

A new moon is when only the shadowed side of the moon is seen.

A waxing *crescent is when the moon is more than half in shadow* and the shadow is receding from the moon.

A first *quarter moon is when the moon is half in shadow* and the shadow is receding.

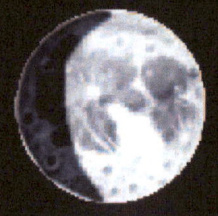
A waxing *gibbous is when the moon is less than half in shadow* and the shadow is receding.

A full moon is when the moon is in the full light of the sun.

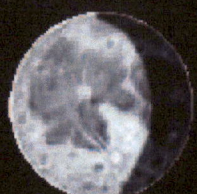
A waning *gibbous is when the moon is less than half in shadow* and the shadow is growing.

A second *quarter moon is when the moon is half in shadow* and the shadow is growing.

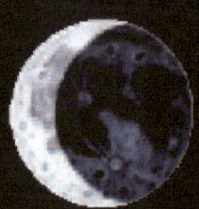
A waning *crescent is when the moon is more than half in shadow* and the shadow is growing.

> Psalm 89: 37 It will be established forever like the moon, the faithful witness in the sky.

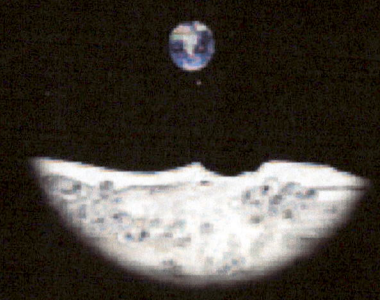

The sun and moon look to be about the same size from Earth, however, the sun is much, much larger. Rarely, the moon will pass directly in front of the sun and block the light. *A solar eclipse is when the Earth is in the moon's shadow.*

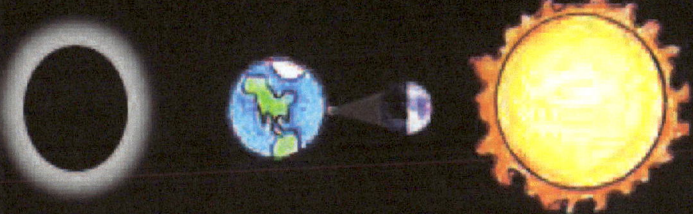

A total solar eclipse means only the sun's corona is visible. The rest of the sun is completely covered. The moon's orbit is elliptical (oval) and slightly tilted which means solar eclipses can only occur once every 18 months or so. Only a small area of earth will see it as a total solar eclipse while other parts will see it as a partial eclipse.

There are accounts in the Bible of solar eclipses. When Christ was crucified, darkness came over the land. Some people believe this means the clouds blocked out the sun, others believe it could mean a solar eclipse. Matthew 27:45 "Now from the sixth hour until the ninth hour, there was darkness over all the land." This was foretold in Amos.

> Amos 8:9 "In that day," declares the Sovereign Lord, "I will make the sun go down at noon and darken the earth in broad daylight.

Also on occasion, the Earth will block the sun's light from reaching a full moon. *A lunar eclipse is when the full moon is in the Earth's shadow.* This occurs at least twice a year. Sometimes the moon will dim in color, other times, in the event of a total lunar eclipse, the moon can look red, brown, orange, or dark yellow. The moon looks like this because of the Earth's atmosphere. The blue, green, and yellow light scatters in the atmosphere while the reds and oranges of the spectrum pass through to the moon. This scattering is also what causes the red and orange colors of a sunset.

Our moon, just like all the other planets and moons, has gravity. Along with the sun, *it pulls on earth's oceans making them rise and fall. This is what gives us tides.* The sun and moon pull on opposite sides of the Earth.

Sometimes, the sun and moon are both pulling from the same side. This is called a spring tide. The high tides are higher and the low tides are lower.

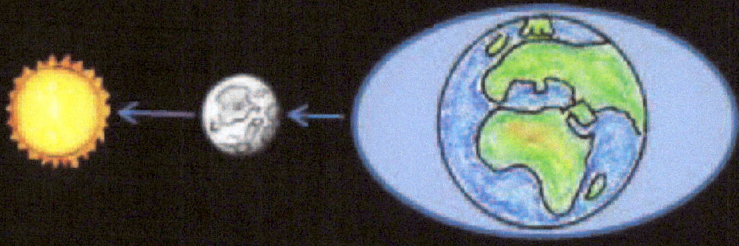

There may also be times when the sun and moon are at right angles from each other. These tides are often very small and less noticeable. This is called a neap tide.

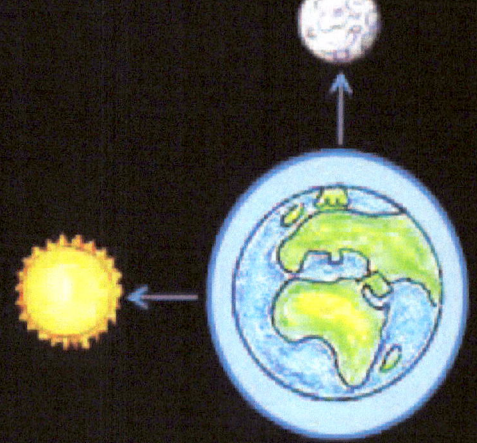

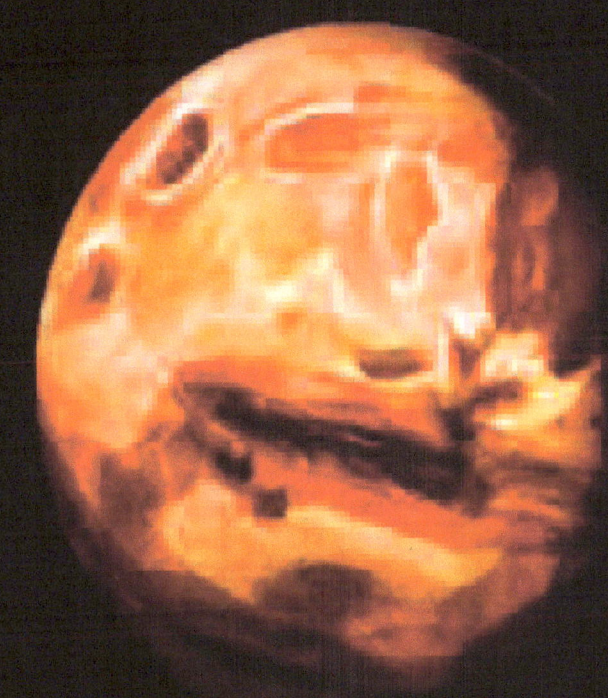

Mars

Mars is the fourth planet from the sun. It has two moons, Deimos and Phobos. Like most moons in the solar system, they are not round but instead look like asteroids. There is no life on Mars nor has there ever been. *Mars is very cold.*

Mars is a large, cold desert covered in lots of red dust. The dust is red due to the iron-oxide in it. Because of this red coloring, the planet is also known as the Red Planet. Mars has a vastly varied terrain somewhat similar to Earth's but on a grander scale.

The fourth planet

Size: 13,263 miles in circumference 4,200 miles in diameter

Distance from sun: 142,000,000 miles

Origin of name: Roman god of war (received this name due to the red coloring)

Moons: 2

Average temperature: -60 ° F

Weight: 38 lbs. (If 100 lbs. on Earth)

There are canyons, mountains, craters, and dead volcanoes covering much of Mars as well as plains. The largest volcano is the extinct Olympus Mons at 15 miles high. There is also a massive canyon called Valles Marineris which is estimated to be 5 miles deep and 1,800 miles long, the same size as the United States from the Pacific to the Atlantic Ocean.

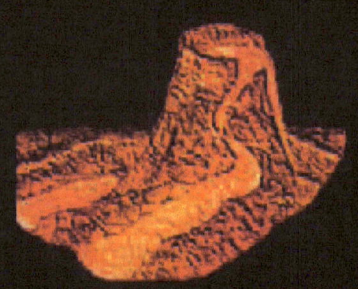

Mars' day is very similar to Earth's. Mars' day is 24½ hours long, and it takes 687 Earth-days to orbit the sun. In addition to having a similar day length, Mars also has 4 seasons and poles covered in ice and carbon dioxide snow. Mars is basically a very cold desert, but in the summer *the temperature can reach a comfortable 70°F. The poles, however, can be as cold as -220°F.* On other areas of the planet, the temperature averages at about -60°F. Mars has a very thin atmosphere of carbon dioxide. It is so thin, it cannot retain much of the sun's heat so when the sun goes down the temperature plummets.

We explore Mars more than any other planet. NASA sends robots called rovers to Mars to take pictures and collect rocks. Already there have been over 20 probes sent to the Red Planet and scientists are researching a way to possibly send astronauts there some day.

Jupiter

Jupiter is the 5th planet from the sun and is also the largest planet in the solar system. Over 1,000 Earth's could fit inside. In fact, if all the planets were put together, Jupiter would still be twice as heavy.

So far, Jupiter has 63 known moons. They range in size from Ganymede at 3,373 miles in diameter to Leda at 10 miles in diameter. Jupiter is so large and has so many moons it is like its own solar system. Jupiter's farthest known moon orbits over 24,000,000 miles away.

The fifth planet

Size: 279,000 miles in circumference 88,846 miles in diameter

Distance from sun: 483,000,000 miles

Origin of name: most powerful of Roman gods

Moons: 63 known

Average temperature: on the cloud tops -230 ° F

Weight: 260 lbs. at cloud top

Jupiter is a giant ball of gas made mostly of hydrogen and helium with small amounts of ammonia and methane. It is the largest and first of the gas giants in the outer solar system. The entire planet is covered in an atmosphere believed to be over 1,000 miles deep. *Some astronomers believe it has a small rock center.* Above this rock center is believed to be liquid and metallic hydrogen. Some believe the atmosphere simply becomes denser and denser the farther in one travels until it becomes liquid and then finally solid. Jupiter also has a very thin set of rings made of dust.

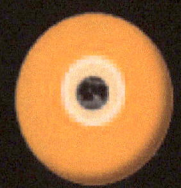

Up in the cloud tops it is very cold. The average temperature in the clouds is −230°F, however deep in the core it is believed to be over 43,000°F. Because of its size and internal temperature, some believe Jupiter to be a failed star or brown dwarf. It is so large it pulls objects into its orbit keeping them from coming near Earth. Perhaps this is one reason God made Jupiter.

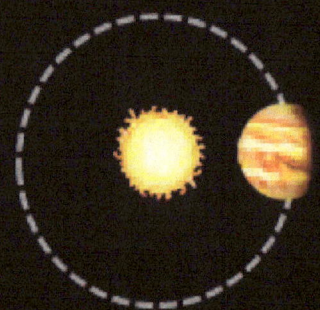

Jupiter's day is about 10 earth-hours long making it the fastest rotating planet in the solar system. It takes 12 years to orbit the sun. Because of Jupiter's rapid rotation, strong winds blow across the planet at up to 300 mph creating storms.

Jupiter has a Great Red Spot that astronomers believe may be a big storm. This storm is believed to be similar to a hurricane and has lasted over 300 years. The spot is about 17,000 miles long and 9,000 miles wide and is so large, two Earth's could fit inside. The spot was first discovered in 1664 by a man named Robert Hooke.

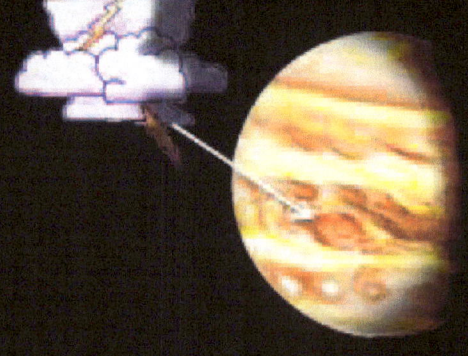

Saturn

Saturn is the 6th planet from the sun and is also known as the Ringed Planet. *It is the second largest planet and is another giant ball of gas.* Similar to Jupiter, its atmosphere is made up mostly of hydrogen and helium with small amounts of methane and ammonia. Astronomers believe it too has a small rocky center surrounded by a liquid and metal core of hydrogen.

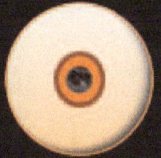

Saturn has at least 60 moons ranging in size form Titan at 3,200 miles across to Pan at 12 miles across and even smaller moons at a mile across. Over half of Saturn's moons orbit more than 6 million miles away. There is even some thought that there may be moons within Saturn's rings.

The sixth planet

Size: 235,000 miles in circumference 74,896 miles in diameter

Distance from sun: 890,000,000 miles

Origin of name: Roman god who is the father of Jupiter- god of agriculture

Moons: 60 known

Average temperature: on the cloud tops -300 ° F

Weight: 110 lbs. at cloud top

It is believed that Saturn is able to float. Based on what is known, Saturn is less dense than water. If there was enough water for Saturn to fit into, it would float.

Saturn is surrounded by beautiful rings made of ice, dust, and rock. What appear to be three broad rings are actually seven or eight rings with gaps in between. The rings are over 180,000 miles wide, but only 40 to 400 feet deep. The rings orbit at different speeds and are believed to contain moons called "shepherd moons". These moons' gravity help hold the rings in place.

Galileo discovered Saturn in 1610. He did not know at the time that Saturn had rings and believed they were big "ears" or "handles".

Saturn's rings are slowly spreading away from the planet. If Saturn was billions of years old, the rings would be gone.

Saturn's day is about 11 hours long, and it takes 30 years to orbit the sun. Because of its rotation speed, the atmosphere is separated into thin bands. Winds near the equator can reach 1,000 mph.

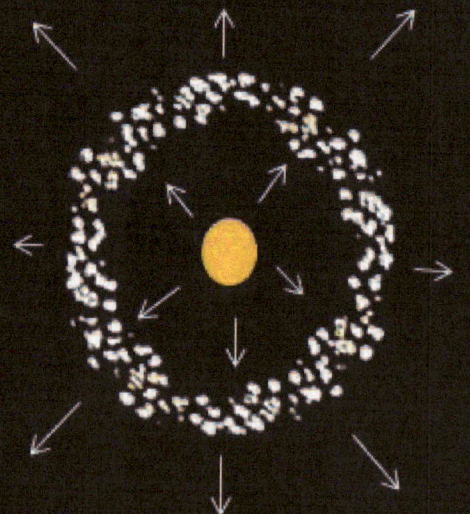

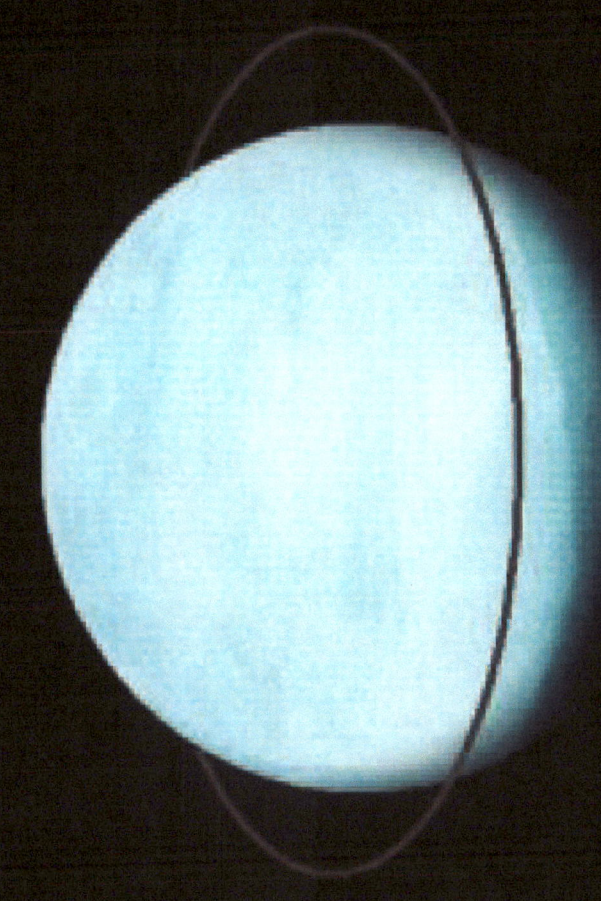

Uranus

The seventh planet

Size: 99,787 miles in circumference 31,763 miles in diameter

Distance from sun: 1,784,000,000 miles

Origin of name: most ancient of Roman gods

Moons: 27 known

Average temperature: average -370 ° F

Weight: 90 lbs. at cloud tops (If 100 lbs. on Earth)

Uranus is the 7th planet from the sun and was discovered in 1781 by William Herschel who originally named it George's Star after King George III.

Uranus has 27 known moons. Of these moons, only five are round, the rest are all asteroid-like in appearance.

Uranus is another planet made mostly of gas. Hydrogen is its main gas followed by helium, and

methane. The methane gives the planet its blue-green color. Below this gas is believed to be a dense, methane, ammonia, and water slush followed by a rock center.

Similar to Jupiter but unlike its size, Uranus has a thin band of 13 *rings.* Unlike Saturn's bright reflective rings, these rings are dark and made up of black dust and rocks only a few inches to 10 yards across. The rings are also just a few miles wide.

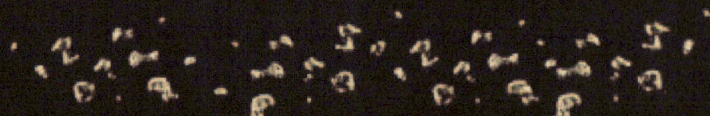

Uranus is unique as it is *the only planet that spins or rotates on its side.* Once again, astronomers cannot explain why it does this, but some have come up with a guess that something large must have crashed into it and pushed it over. *It does this because God made it this way.* Proof, yet again, that God created the heavens.

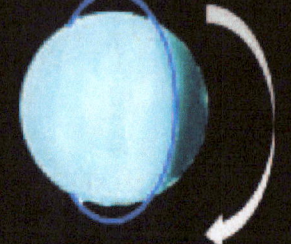

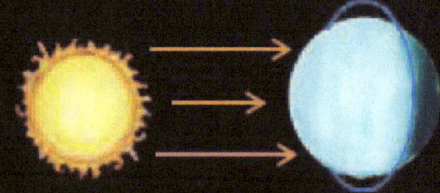

It takes about 84 earth years to orbit the sun and 17 hours to complete a rotation. *Because of its tilt, the poles each spend half the year in day and the other half in night. That's 42 years!*

Uranus is the coldest planet in the solar system. It is around -370°F and is called the "Ice Giant". The reason Uranus is so cold is because its core is so cold. It is only about 8,500°F. Scientists do not know why it is this cold but believe it has something to do with the planet's sideways rotation.

Neptune

Neptune is the 8th planet from the sun and also known as Uranus' twin. The two planets are very similar in composition and size.

Neptune was discovered in 1846 by four astronomers in England and France when they noticed Uranus was not orbiting like they thought it should. They did some measurements and discovered Uranus.

Neptune has 13 known moons. Triton, the largest moon at 1,681 miles across, orbits in the opposite direction of all the other moons.

The eighth planet

Size: 99,786 miles in circumference 30,775 miles in diameter

Distance from sun: 2,795,000,000 miles

Origin of name: Roman god of the ocean

Moons: 13 known

Average temperature: average -330 ° F

Weight: 110 lbs. (If 100 lbs. on Earth)

Naiad is Neptune's smallest moon at only 36 miles wide.

Neptune is the last of the gas planets. Just like Uranus, its atmosphere is mostly hydrogen with some methane. The planet is blue due to the methane gas. The difference in color between Uranus and Neptune is the amount of methane in the atmosphere. Below this atmosphere is believed to be an ocean of water, ammonia, and methane with a rock center.

Neptune is also the windiest planet in the solar system with sustained winds up 1,300 mph.

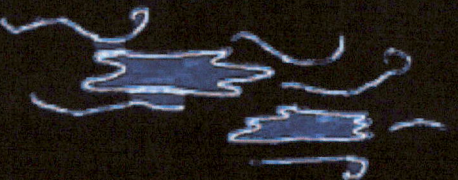

Like Jupiter, Uranus often has storms appear as large spots in the atmosphere. *Neptune once had a large spot called the Big Dark Spot.* It too was believed to be a massive storm. However, unlike Jupiter whose storm has lasted hundreds of years, Neptune's storms have since faded.

Neptune is the farthest planet from the sun and is very cold. The temperature remains pretty consistent (it does not change much) *at about -300°F.*

Neptune takes almost 165 years to orbit the sun. Since its discovery in 1846, Neptune has only just completed its orbit. *Its day is 16 hours long.*

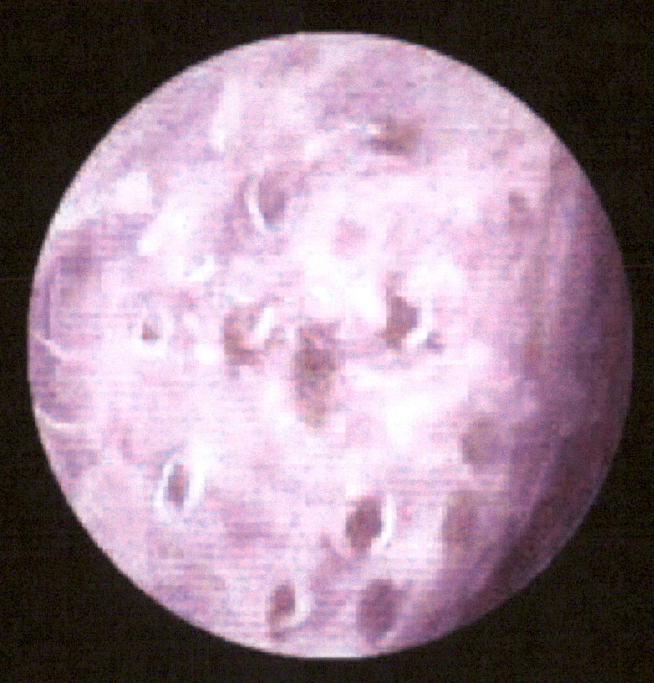

Pluto
and dwarf planets

When Pluto was first discovered in 1930, it was considered the ninth planet. For years many scientists disagreed that Pluto was a planet. It was when Eris was discovered in 2003 that Pluto's classification was reconsidered. In 2006, it was downgraded to dwarf planet making Neptune the last planet in the solar system. *Pluto is called a dwarf planet because it is so small. It is smaller than our moon* as well as several other moons in the solar system.

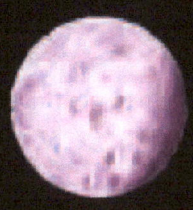 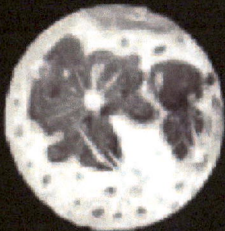

Pluto is only 1,400 miles in diameter, or 4,500 miles in circumference. Despite having a round shape, *Pluto is not a planet like others because it has not cleared its path (orbit) of material.* It is large enough to have a

The dwarf planet

Size: 4,493 miles in circumference 1,432 miles in diameter

Distance from sun: 3,675,000,000 miles

Origin of name: Roman god of the underworld

Moons: 3 known

Average temperature: average -350 ° F

Weight: 6.5 lbs. (On Pluto, if 100 lbs. on Earth.)

round shape, but not large enough to clear its path. Within Pluto's orbit is other space material, asteroids, meteoroids, and so on. Objects with these characteristics are called dwarf planets.

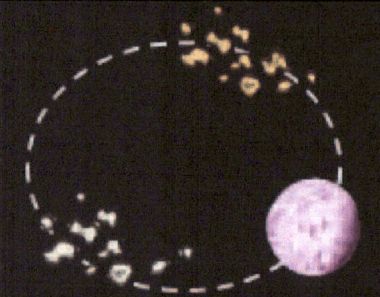

Pluto also doesn't orbit like the other planets. *Sometimes Pluto crosses over Neptune's orbit.* It also orbits on a different plane than the other planets in relation to the sun.

Pluto is very far from the sun, and is so cold it is frozen ice and rock. The thin atmosphere actually freezes onto the planet and only appears when Pluto is closer to the sun and begins to thaw. The surface is a thin crust of methane ice.

Pluto has three moons. Charon is the largest and is so big that Pluto and the moon orbit each other. This is called a binary system. Charon and Pluto orbit around a point between the two bodies. They also keep the same sides facing each other as they orbit. This system was discovered in 1978 when an astronomer noticed a bulge appearing sporadically beside Pluto.

Pluto takes 248 years to orbit the sun and just over 6 days to rotate once on its axis.

There are other dwarf planets in our solar system. In the asteroid belt between Mars and Jupiter lies Ceres at about 580 miles across. Beyond Pluto in another asteroid belt known as the Kuiper Belt lies Eris. There is some debate whether Eris or Pluto is bigger. Either way, Eris and Pluto are the two largest known dwarf planets.

Chapter 4

Stars, Galaxies, Nebulae, and the Sun

Stars

Stars are huge balls of hot gas. The gas fuses and creates great amounts of energy that *give off heat and light.* Stars are mostly all hydrogen and have enough fuel to burn for billions of years.

Stars were created in different colors. Stars are different colors because of their temperature.

Stars can be red, orange, yellow, white, or blue.

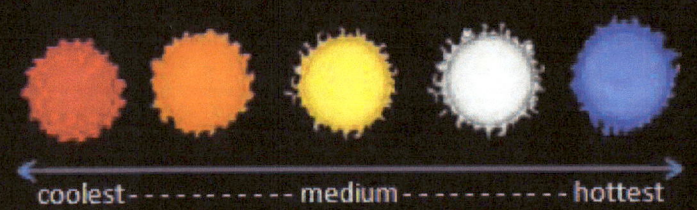

Normally people think of red when thinking of something extremely hot, however, a *blue star is actually the hottest and a red star is the coolest.* A red star is a "cool" 5,800°F. While a blue star is over 17,500°F. The other star temperatures fall somewhere in between. Our sun is about 9,900°F

Stars can even change colors. These are called pulsating stars. They are giant stars that continually grow and shrink changing their color and surface temperature. As the star expands and moves away from the hot core, the surface cools and begins to redden, when it shrinks back towards the core, the surface heats up and changes color towards the whites and blues. Our sun is actually a very consistent, stable star. If we had a pulsating star in the center of our solar system, life could not survive.

Stars also change color and temperature depending on what stage of life they are in. As they burn off their gas, the star begins to expand, cooling. If the star expands to a red giant, often the gas will simply disperse around the star as a planetary nebula, leaving a white dwarf.

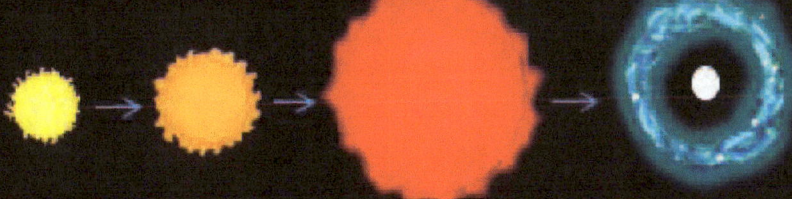

If the star expands to a super red giant, it will often explode as a supernova.

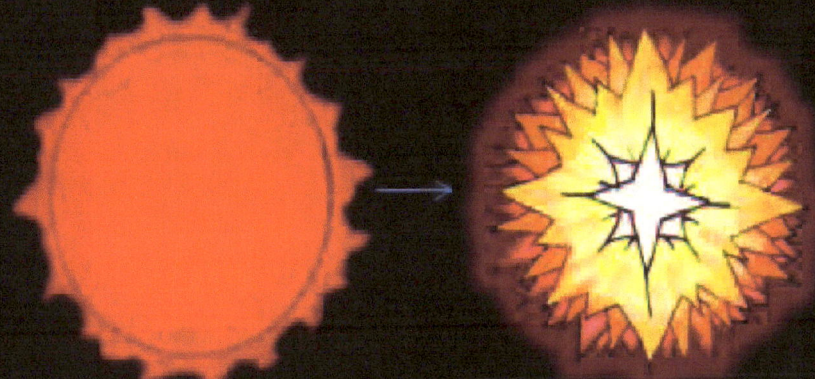

This may also leave behind a neutron star, (an extremely dense star made entirely of neutrons), or the nebula may just float away as the hot gas continues to expand. The third option is that the dying star becomes a black hole.

Stars are often found in groups or clusters. When they cluster together to make a picture it is called a constellation. For thousands of years, people have used these as *maps* to help them know which way is North for navigation.

Here are a few constellations we can see in the northern hemisphere throughout different times of the year:

The Little Dipper

The top most right star of the Little Dipper is the North Star. The North Star does not change positions in the night sky while all other stars and constellations seem to rotate around it. Because of this, it is used often as a navigating star.

Orion (The hunter)

Pegasus (The winged horse)

Ursa Major (The Great Bear)

Pisces (The fish)

Draco (The dragon)

Leo (The lion)

Cygnus (The swan)

Occasionally, what may appear to be a really bright star may actually be a binary star. A binary star is when one star orbits another.

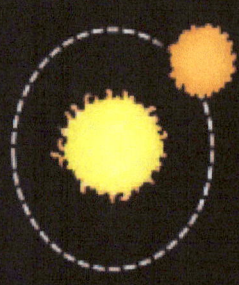

Stars twinkle because of the earth's atmosphere. The star light comes to Earth in a straight line, but then t*he light bends* from the moving air of our atmosphere. This is called stellar scintillation. The more atmosphere the light has to travel through, the more it twinkles. So, stars on the horizon twinkle more than those directly overhead. The reflected light from the planets also comes to us in a straight line. The reason planets do not twinkle, is not because they are planets, but because they are so close (relatively speaking). Occasionally, however, if it is a very windy day, even planets will twinkle.

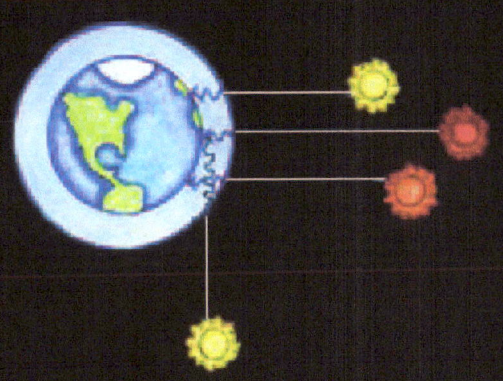

A falling star is not really a star at all. It is a meteoroid burning up as it enters the Earth's atmosphere.

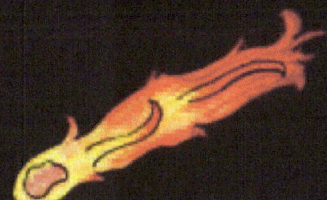

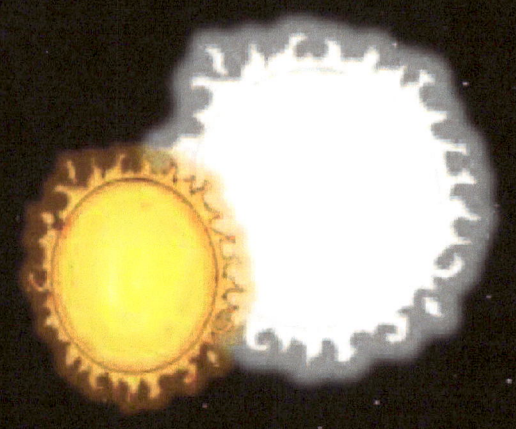

The Sun

Our sun is a star. It is close enough that we can feel its heat, but it is 93,000,000 miles away.

God gave us the sun for light and heat. Plants need the sun to make food. We need plants for food and air. Without the sun, Earth would be too cold for anything to live. Plants, animals and man could not survive.

 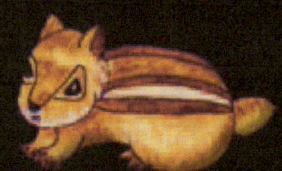

The visible light we see is called the spectrum. It is red, orange, yellow, green, blue, indigo, and violet. All stars emit these colors. When all of these colors are  present in equal amounts, the light will be white. Some stars emit more of one color over another due to its temperature. *Our sun is technically a yellow star* because it emits slightly more light in the yellow spectrum. However, because the sun is so close, and so bright, we cannot see this very slight difference. (Do not look directly at the sun!)

We see all of the colors, so the sun is white. We can see the sun's white coloring in space and occasionally here on earth like in the sun's reflection in water or windows. For the most part, here on earth, our sun appears yellow. This is because of the earth's atmosphere. The atmosphere scatters some of the blue light (which is why the sky is blue), making the sun appear more yellow.

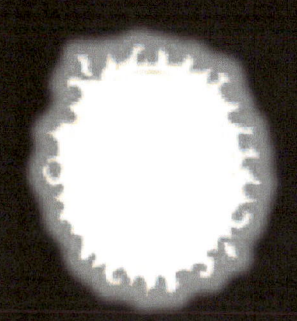

Sometimes we can see the individual colors that make up sunlight. When the sun shines through a prism or a cut piece of glass, a rainbow may appear. Each color of the rainbow has a different wave length. As each wave length enters the prism at an angle, it travels at different speeds, so that each color is visible. This is how rainbows form when it rains. The tiny water droplets in the air separate the sun's light so that each color becomes visible. We can also see this when the sun sets. Many colors can be seen because of the sun's light traveling farther through the atmosphere.

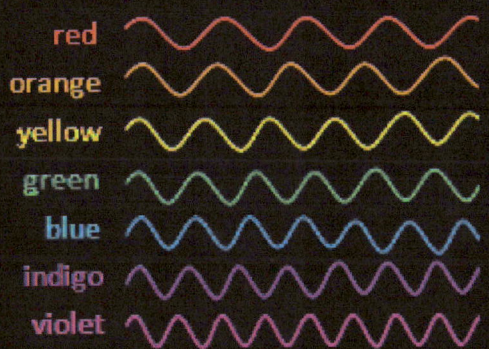

Occasionally, we can see other colors in our atmosphere. This is not because of the sun's visible light but because of the sun's solar wind. Solar wind is a stream of *tiny charged particles flying off the sun into space. The Earth is like a giant magnet, and it is surrounded by an ivisible magnetic field.* This field enters and exits at the poles and protects us from some of the sun's harmful rays by deflecting them into space.

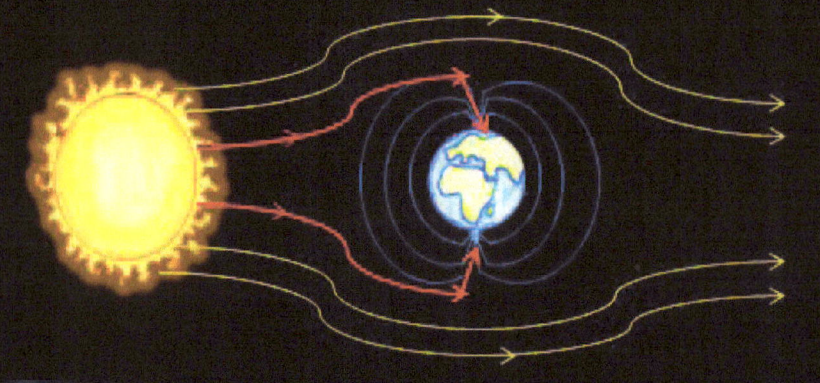

Sometimes the charged particles travel down the top and bottom of the magnetic field. When the charged particles touch the gases in our atmosphere, they cause the air to glow and make beautiful colors. This is called an aurora. *The aurora can be green, red, blue, or purple. It depends on how high in the atmosphere the particles are and also with what gas the* particles react. At higher altitudes, the oxygen creates a red aurora, at lower altitudes, it is bluish-green. Nitrogen at lower altitudes produces a red aurora. Blue and purple auroras are produced from helium and hydrogen, though these colors are more difficult to see.

Auroras in the northern hemisphere are called aurora borealis. In the southern hemisphere, they are called aurora australis.

Our sun sometimes has dark marks on it. These are called *sun spots.* Sun spots form when a loop from the sun's magnetic field brings less dense, cooler gas to the surface. The spots always appear in pairs where the magnetic line exits and reenters the sun. The cooler gas is about 6,300°F and appears dark.

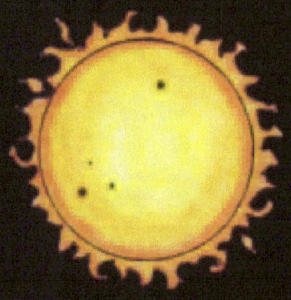

Sometimes the sun will have huge flames. These are called *solar flares.* Super-hot gas is thrown out into space for thousands of miles. They can last for several days, disrupt satellites and other communications here on Earth, and can even cause power outages. Solar flares are slightly different from solar prominences. A solar prominence occurs when the sun's magnetic field loops up through the surface carrying cooler gas and forms an arch.

The sun is almost completely hydrogen with some helium. In the core, hydrogen is fused into helium from 27,000,000°F of intense heat and pressure. The energy radiates out through the radiative zone and cycles around in the convective zone before finally surfacing as heat and light in the photosphere. The photosphere is the part of the sun that we see and is the beginning of the sun's atmosphere. It is around 210 miles wide and almost 10,000°F. In the chromosphere (the second layer of atmosphere), the heat is magnified to 17,500°F. The corona is the final layer of the atmosphere and is virtually invisible, though it can be seen during a solar eclipse.

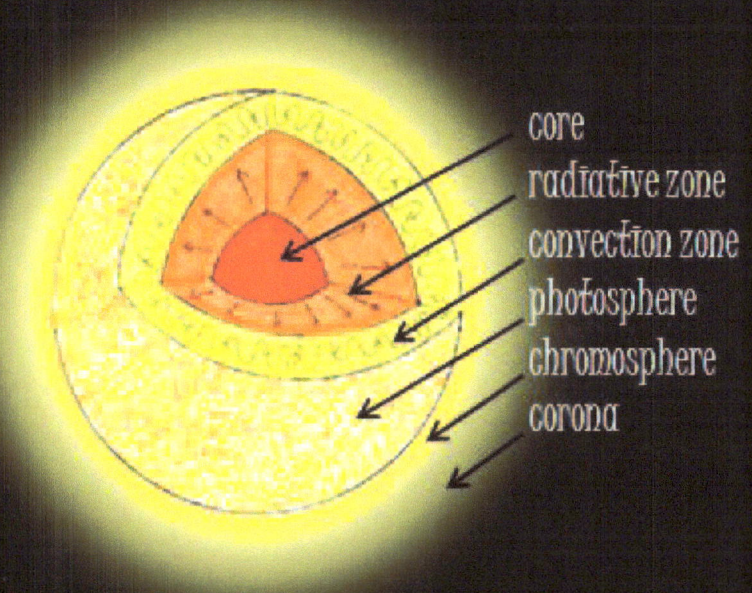

core
radiative zone
convection zone
photosphere
chromosphere
corona

(Remember; don't look directly at the sun!) The corona extends millions of miles out into space. In this part of the sun's atmosphere, the sun's heat is magnified again to an extremely hot 3,600,000°F! No one knows for sure how the heat is intensified in this area, but some believe it is because of the sun's magnetic field. The corona also hurls charged solar particles into space as solar wind. This whole process takes 200,000 years! Every second, four million tons of fuel is burned.

The photosphere is not smooth. It is covered in granules. Convection currents of plasma in the convective zone bring the hot plasma to the surface. The plasma releases energy, cools, and sinks. It is similar to bubbles in boiling water.

Stellar Spectroscopy

400	450	500	550	600	650	700
indigo	blue	blue-green	green	yellow-orange	dark	red

Astronomers can determine of what gases a star is composed by using a device called a spectroscope. A spectroscope is basically three items: a slit for the light to pass through, a lens, and a prism. When an element is heated it emits a unique "fingerprint" of light that is visible in the spectrum.

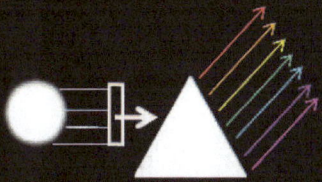

Each element has unique absorption lines (where the element absorbs light) that appear as black lines on the spectrum. When viewing star light through the spectroscope, astronomers can tell exactly what elements and gases the star is made based on where these lines occur in the spectrum. Below is an example of the absorption lines of hydrogen.

Because the absorption lines absorb light, the emission lines are just the opposite. They emit light. The emission lines are unique to the element and appear in the exact location of the black absorption lines, only they are in color and the rest of the spectrum is darkened. As seen below, the hydrogen spectrum shows the emission lines in the same area as the absorption lines.

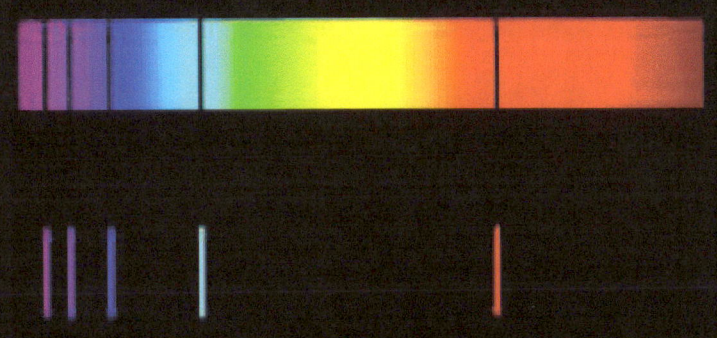

If these lines appear in the exact configuration (line up) but are shifted towards the red end of the spectrum, astronomers know the star is moving away from the Earth. (Remember red is cooler.) If the lines are shifted towards the indigo end of the spectrum, astronomers know the star (or galaxy, or nebula) is moving towards the Earth. (Remember that blue is hotter than red.) If the lines are exactly where they should be, the celestial body is staying at the same distance from earth. This shifting is called the Doppler Shift.

Galaxies

Our star and solar system are just a tiny part of a massive galaxy called the Milky Way. In the universe, there are several different types of galaxies.

The _barred galaxy_ has a long bar-like center with two arms coming out from either side. Some astronomers believe the Milky Way to be this type of galaxy.

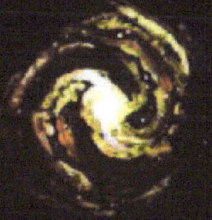

The _spiral galaxy_ has a bulging disk in the center with arms spiraling out from the middle. It typically has two main arms with several smaller arms. Most astronomers believe the Milky Way to be a spiral galaxy.

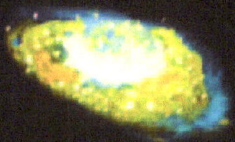

An _elliptical galaxy_ is a group of stars clustered together. It looks like the center bulge of a spiral galaxy.

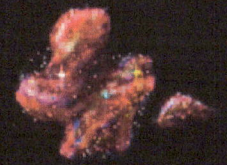

An _irregular galaxy_ is a group of stars and dust with no particular shape.

A _lenticular galaxy_ is a group of stars scattered along a line.

light-year

Our solar system encompasses billions of miles. Given that the distance around our planet is 24,000 miles, the idea of millions of miles is barely conceivable, let alone billions. Once outside the solar system, the distance to the next celestial object is so vast, it is measured in light-years. A light-year is the distance light can travel in one year. Light travels at a speed of 186,000 miles per second. Multiply that distance by seconds in a minute by 60 minutes in an hour by 24 hours in a day (186,000 miles x 60 seconds x 60 minutes x 24 hours) and the answer is 16,070,400,000 miles in one day! Multiply that by days in a year (x365) and the number is 5,865,696,000,000 miles a year! Then, multiply that by the millions and billions of miles separating us from other galaxies, and the distance is mindboggling! How awesome and powerful is our God! It's easy to see why this method of measurement was chosen!

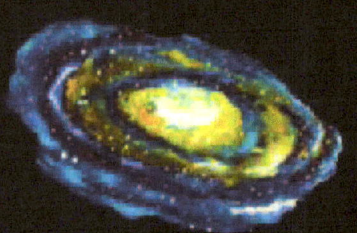

The Andromeda Spiral is the nearest spiral galaxy to the Milky Way. It is about 2.5 million light-years away. Imagine multiplying 2,500,000 by 5,865,696,000,000 and having to write that number whenever the distance to Andromeda is addressed! Andromeda is actually over 1 quintillion miles away! That's 1,466,424,000,000,000,000!

Nebula

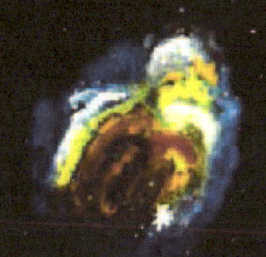

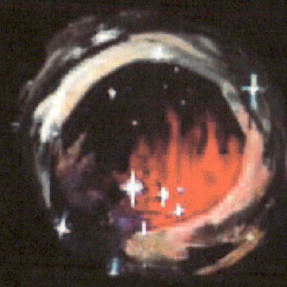

Besides stars and galaxies, the universe is also home to nebulae. *A nebula is a cloud of dust and gas.* It can form after the death of a star, or simply just exist floating in space. Hydrogen and helium are the main gases within these nebulas. *If a nebula is very close to a hot star, the solar radiation causes it to glow. Other times, it may glow from reflecting the light of stars. The beautiful colors come from the gases within*, how the nebula was formed, or the reflection of light. Other colors appear from different elements inside the nebula.

> Deuteronomy 3:24 Sovereign Lord, You have begun to show to Your servant Your greatness and Your strong hand. For what god is there in heaven or on earth who can do the deeds and mighty works You do?

There are five types of Nebulae.

<u>Dark nebulae</u>: These are dark clouds of gas, virtually invisible unless silhouetted by stars.

Emission nebulae: These are brightly lit due to the high temperature of the gas and the ultraviolet radiation from a nearby star.

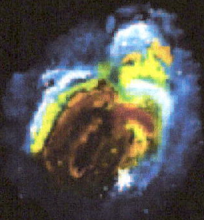

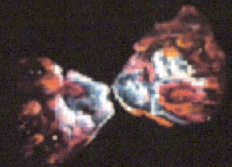

Planetary nebulae: These form from a dying star. They are lit from within by a white dwarf. This planetary nebula is called a butterfly nebula.

Reflection nebulae: These do not emit light. They glow by reflecting the light of nearby stars.

Supernova remnants: These are the remains of a supernova. They glow from within from a white dwarf or a neutron star.

*nebulae are the plural form of nebula

Chapter 5

Man has been observing the skies for thousands of years. Ancient peoples traced the movement of the stars and planets and followed the phases of the moon. They observed the sun as it traveled across the sky and drifted across the horizon where it rose and set throughout the year. They were curious about what they saw and drew pictures and wrote stories about what it meant. Unfortunately, some of these ancient peoples worshipped the sun, moon, stars, and planets as gods. The Egyptians, for example, worshipped the sun, and the Aztecs worshipped Venus. They worshipped what God created instead of Him. As Romans 1:25 says, "They exchanged the truth of God for a lie, and worshipped created things rather than the Creator."

Sadly, some people today look to the stars for guidance. This is called astrology. People who practice astrology believe the stars, planets, and universe, control their future. But the Bible says *God made the stars to show how powerful and majestic He is.* "For since the creation of the world, God's invisible qualities- His eternal power and divine nature- have been clearly seen, being understood from what has been made, so that men are without excuse." Romans 1:20. The universe was created for His glory, and He knows every detail of it. *We can look at the stars at night and remember that He made them and gave each one a name!*

> "He determines the number of the stars and calls them each by name." Psalm 147:4

A few hundred years after Christ, a Greek astronomer named Claudius Ptolemy came up with a theory that the Earth was the center of the solar system and that all celestial objects revolved around it. This is called a geocentric model. Ptolemy was a brilliant man and came up with many theories and formulas that are still used today. His model of the earth-centered solar system was more complex than the planets revolving on a flat plane. The orbital path also rotated around the earth. It was very close to copying the movement of the planets. It was so close in fact, it was believed to be true for over 1,400 years.

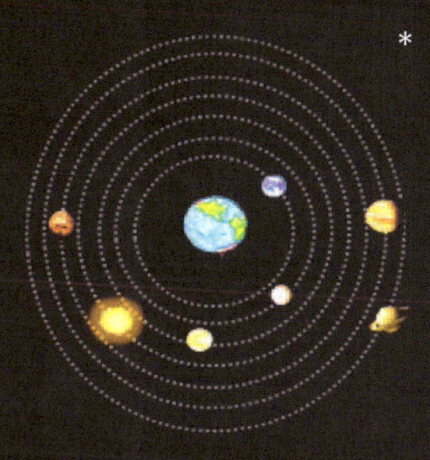

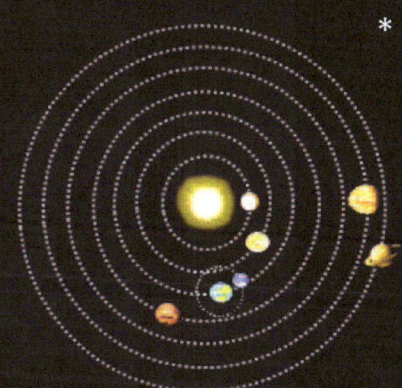

It wasn't until the late 1400's/ early 1500's that Nicolaus Copernicus introduced the theory that the Earth and planets orbited the sun. This is called a heliocentric model. Unfortunately, even though he was correct, he did not have the proof needed to back his theory and many people did not believe him. It was several decades before Johannes Kepler supported Copernicus with his laws of planetary movement.

Another man, Tycho Brahe, born after Copernicus' death, provided yet another theory that the planets orbited the sun, but the sun orbited the Earth. There were some pretty complicated diagrams as astronomers attempted to explain their incorrect theory that everything orbited the Earth.

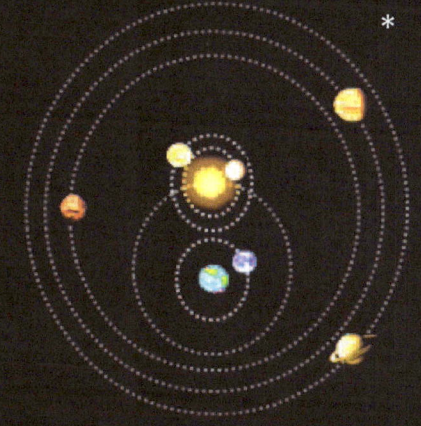

* Solar systems are shown without Uranus and Neptune as the first models would have been.

The 1600's, the Telescope, and Gravity

For thousands of years, man had only his eyes to view the stars and planets. Then, in 1608, a Dutch man named Hans Lippershey invented the telescope. He made glasses (lenses) for a living. It is said that he created it when he saw his children playing with the lenses he brought home. *This first telescope was a refracting telescope that used lenses to focus the light.*
Most likely, Lippershey did not use his new invention to look at the heavens. Galileo heard of this new invention, built one of his own, and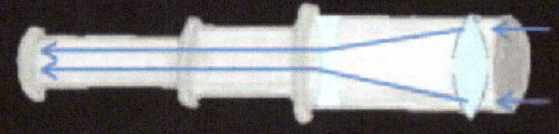
is credited for first using the telescope for astronomy. The light first enters through an opening called the apature. It is then focused with a convex lense to a second concave lense which then magnifies the light.

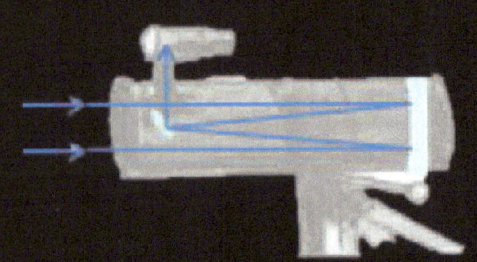
Later, Sir Isaac Newton invented the reflecting telescope. It uses mirrors to reflect and magnify the light. Mirros are significantly cheaper to make and can be made much larger than lenses allowing the user to see farther into space. *Some of the first telescopes could see as far as Saturn. Now, some telescopes can see billions of light-years away.*

It is said that in 1665, while studying Kepler's theories about the planets, Sir Isaac was staring at the moon while sitting under an apple tree when an apple fell on his head. He realized the moon and all the planets were held in place by gravity. The Earth's gravity held the moon in orbit while the sun held the planets in orbit.

With Nicolaus Copernicus' theory of a sun-centered solar system, the invention of the telescope, Johannes Kepler's accurate theory of planetary movement, and Isaac Newton's scientific contributions on gravity astronomers were able to better understand and accept our sun-centered solar system. It also brought about the discovery of Uranus and Neptune as well as nebulae. The telescope opened the heavens to man and gave a more detailed view of the Milky Way.

1940's - 1960's The Space Race and the Lunar Landing

Since the creation of gunpowder, rockets have been used as a weapon. It wasn't until October 3, 1942, that Germany first used a rocket to reach into space. The rocket soared some 50 miles above the surface. Following that, *in the 1950's, more countries started shooting rockets into space. The rockets had to be very powerful to get into space.* Then, on October 4, 1957 the Soviet Union (of what Russia was once a part) launched Sputnik. It was the first satellite to be launched into space and orbited the earth every 98 minutes. Other *satellites were launched which could take pictures and send signals back to earth.* This began what became known as the Space Race. It was a time when the United States was in a race to beat the U.S.S.R. (Soviet Union) with our space program. NASA (National Aeronautics and Space Program) was created just less than one year after Sputnik was launched. In the 1960's our president at the time, President John F. Kennedy, challenged that by the end of the decade, America (NASA) would put a man on the moon and bring him back safely. The USA and USSR were now in a race to the moon as both countries sought to be first.

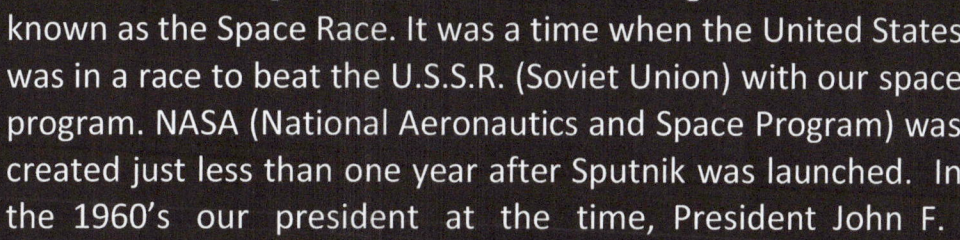
Sputnik

In order to be sure a space craft was safe for people, both the Americans and the Soviets sent animals into space, a monkey and a dog respectively. *Then, on April 12, 1961, the Soviets sent the first man,* Yuri Gagarin, *into space*. Less than a month later on May 5, 1961 *Alan Shepard became the first American in space. Both men had to wear special suits because there is no air in space. People who go into space* (in America) are *called astronauts.* This comes from the Greek words astron (star) and nautes (sailor). So an astronaut is a star sailor! A Russian astronaut is called a cosmonaut which is from the Greek words kosmos (universe) and nautes (sailor). So, cosmonaut means universe sailor!

It was another eight years before America was ready to put a man on the moon. The astronauts had to undergo training to prepare them to fly a rocket in space as well as to operate the equipment on the moon. NASA first developed the Lunar Orbiter program in 1966-1967 to orbit the moon, take pictures, and find a safe place for the lunar landing.

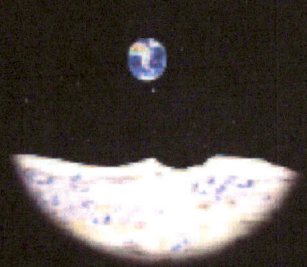

Then on July 16, 1969, Apollo 11 launched from Kennedy Space Center in Florida with three men on board. Apollo 11 was made up of a rocket called the *Saturn V*, the command module called *Columbia,* the service module, and the lunar module called the *Eagle*. It took three days to reach the moon.

Shortly after take-off, after burning up its fuel, the first section of the *Saturn V* fell away.

Then, when it too was empty, the second section fell away.

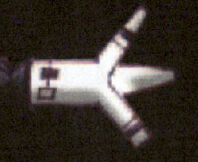 

After this, the command and service module separated from the third section. It then turned around and pulled the lunar module from within the third section.

Eagle

The astronauts began their orbit around the moon on July 19th and searched for a safe place to land. The *Eagle* (lunar module) detached from the command module and headed towards the moon's surface while the command module stayed in lunar orbit. Finally, *on July 20, 1969, two men landed on the moon. They were Neil Armstrong and "Buzz" Aldrin. Another man, Michael Collins, stayed on the command module (Columbia).*

command module: *Columbia* contained navigation instruments; astronauts worked here

Columbia

service module: contained power and life-support

When the men landed on the moon, some people were afraid there would be billions of years' worth of moon dust and the Eagle would sink. Instead, there was only a little which proved God made it not too long ago! You can still see the foot prints from the first steps on the moon.

Neil Armstrong was first out of the *Eagle* and the first person to set foot on the moon. His first words spoken on the moon were: "That's one small step for man; one giant leap for man-kind."

The men walked around exploring the moon and used a lunar rover for transportation. The moon had many craters, rocks, mountains, and basins (flat areas) and was also covered in a fine powdery dust. It was different walking on the moon than on earth because there was less gravity. A 200 pound man would only weigh 33 pounds on the moon. The men felt almost weightless while they hopped around exploring.

The astronauts had other work to do on the moon besides exploring and spent the next 22 hours doing just that. Scientists on earth wanted to study the moon so the astronauts experimented. One experiment tested the solar wind from the sun on different types of material. Another used a seismometer to determine what was inside the moon and test for moonquakes. Armstrong and Aldrin also *collected rocks and took lots of pictures.* They set up an American flag and even talked with President Nixon. One of the most important things they did (which is still in use today) was place a 2-foot wide panel with an array of 100 mirrors. It is called the Lunar Laser Retroreflector. Scientists on earth can shoot a laser all the way to the moon to accurately calculate the moon's distance from earth. (This is how they know it is moving away from us.)

After Aldrin and Armstrong finished exploring and completed the experiments, they climbed on board the *Eagle* and lifted off the moon meeting up with the command module *Columbia* and Collins. When Aldrin and Armstrong were safely on board, they released the lunar module and headed home to Earth. *When it was time to land, the Apollo 11 command module splashed down in the Pacific Ocean.* It was July 24th. The entire trip to the moon and back lasted eight days.

Today

Today we can see far out into space with giant telescopes. Some are on satellites orbiting the Earth such as the Hubble telescope. *Some are built inside a large building called an observatory.* In order to see as far into space as possible, observatories need to be where it is very dark at night. Most are high up on a mountain or in the desert away from the bright lights of large cities.

We also have something called a radio telescope. Similar to the light waves we see as color, stars emit invisible waves called radio waves. Each element also has unique atomic lines in its radio spectrum. A radio telescope detects these radio waves and the radiation emitted from stars, galaxies, and other celestial objects and can determine of what the star is made. Unlike telescopes, they do not have to worry about being in a dark location. Because radio waves are so large, (Some almost a mile long!) more than one radio telescope is often used and is set up in what is called an array.

In the beginning of the space program, NASA used rockets to launch astronauts into space. After landing on the moon or simply going up into space, the majority of the space craft burned up in the atmosphere or was left in space never to be used again. NASA still uses rockets to send astronauts into space but now has a reusable ship. *Today there is a new space craft called the space shuttle that can be used again and again.* It has two rocket boosters which are shed two minutes after launch. They fall back into the ocean and are collected to be used again. The shuttle also has a large fuel tank that is released in space and burns up in our atmosphere. *The shuttle takes off like a rocket, but lands like an airplane.* After it lands, it often has to be transported back to a storage area or hanger.

It cannot take off without its rockets or fly like an airplane in our skies, so a huge 747 airplane carries it on its back.

Machines that look like satellites are sometimes launched into space to explore planets and take pictures. They are called probes. They can explore all kinds of celestial objects including asteroids, planets, and other moons in the solar system. They retrieve other important data about the object including determining the makeup of the object. Then, signals are sent back to Earth with the information they find.

There is also a place in space where astronauts can live, work, and do experiments. It is called a space station. There have been many space stations over the years. The space station most used today is the ISS- International Space Station. There are 16 countries involved in the ISS including the United States, Russia, Europe, and Japan.

Index:

A
asteroid, 3, 4, 8, 9, 31, 37, 42
astrology, 3, 60
astronomy, 3, 60, 62
astronaut, 25, 32, 63-66
B
binary star, 46
C
comets, 3, 10, 11
constellations, 3, 46
D
dwarf planets, 9, 41, 42
E
Earth, 18-24 (Earth is mentioned innumerable times throughout the text. The pages listed are where Earth is discussed exclusively.)
G
galaxy, 54-56
J
Jupiter, 8, 25, 33-35, 38, 40, 42
L
light-year 56, 62
lunar eclipse, 29
M
magnetic field, 4, 49-52
Mars, 8, 31-32, 42

M (cont.)
Mercury, 14-15, 27
meteor, 8-9
meteorite, 8-10
meteoroid, 8-9, 42, 47
Moon (Earth's), 25-30
N
nebula, 45, 54, 57-58, 62
Neptune, 9, 39-42, 61, 62
O
observatory 67
P
Pluto, 9, 14, 41-42
S
satellites, 51, 63, 67, 68
Saturn, 35-36, 38, 62, 64
solar eclipse, 29, 51
stars, 3, 44-48, 55, 57-58, 60, 62, 67
stellar scintillation, 47
stellar spectroscopy, 53-54
sun, 48-52
T
tides, 5, 30
U
Uranus, 37-40, 61, 62
V
Venus, 16-17, 60

Resources

Books
(These are not Christian resources and some will often speak of the Big Bang and in terms of billions of years)

Children's Atlas of the Universe by Robert Burnham
Published by Weldon Own Pty LTD: 2000
ISBN: 978-1-74089-615-3

First Space Encyclopedia by Caroline Bingham
Published by DK Publishers, New York: 2008
ISBN: 978-0-75663-366-0

Inside Space: Comets and Meteors by Jane Kelley
Published by South Pacific Press, USA: 2009
ISBN: 978-1-59198-707-9

Inside Science: Moon and Tides by Rob Lang
Published by South Pacific Press, USA: 2009
ISBN: 978-1-59198-708-6

Life on a Space Station (Discovery Education Stage 3 reader) by Andrew Einspruch Published by Weldon Owen Publishing Pty LTP; Sydney, Australia: 2011
ISBN: 978-1-74252-184-8

There's No Place like Space: All About our Solar System by Tish Rabe Published by Random House, INC; New York: 1999 ISBN: 0-679-99115-8

Voyage across the Cosmos by Giles Sparrow
Published by Quercus Publishing Plc. London, England: 2008 ISBN: 978-1-84724-524-3

Christian Astronomy Books

Destination: Moon by Astronaut James Irwin
Published by Master Books, San Antonio, Texas: 1989, 2004 ISBN: 1-929241-98-4

The Astronomy Book by Dr. Jonathan Henry
Published by Master Books, San Antonio, Texas: 2005
ISBN: 0-89051-250-7

Taking Back Astronomy: The Heavens Declare Creation and Science Confirms It by Jason Lisle
Published by Master Books, San Antonio, Texas: 2006
ISBN-13: 978-0-89051-471-9

The Heavens Proclaim His Glory by Lisa Stilwell
Published by Thomas Nelson, Inc. Wheaton, Il: 2010
ISBN-13: 978-1-4041-8958-4

Websites

A website about lunar eclipses and legends regarding lunar eclipses:
   http://starryskies.com/The_sky/events/lunar-2003/eclipse-Nov8.html

An astronomical calendar describing when and where certain celestial objects can be best viewed:
   http://www.seasky.org/astronomy/astronomy-calendar

Information on the world's largest crater:
   http://www.southafrica.info/about/geography/vredefort-080605.htm

Information about comets: http://earthsci.org/fossils/space/comets/comet.html

Information about the solar system: http://solarsystem.nasa.gov/planets/index.cfm
www.nationalgeographic.com   www.universetoday.com   www.kidsastronomy.com

a model of Ptolemy's geocentric solar system (video) http://planetfacts.org/claudius-ptolemy/

www.ingramcontent.com/pod-product-compliance
Lightning Source LLC
Chambersburg PA
CBHW051026180526
45172CB00002B/486